Time Wave Exploration

John Smedley

Time Wave Exploration: The Intersections of Fractal Time, Quantum Mechanics, and Consciousness

Author: John Smedley

Date: \today

Table of Contents

Introduction: Understanding the Time Wave Theory..........3
The Foundations of Time in Modern Physics 3
Quantum Mechanics and the Illusion of Linear Time..........3
Free Energy and Nikola Tesla's Vision 3
Warp Bubbles and the Alcubierre Drive 3
Simulation Theory and the Nature of Reality..........3
The Mandela Effect and Evidence of Temporal Shifts 3
Bridging Consciousness, Time Waves, and Quantum Mechanics 3
Future Research and Implications for Humanity..........3
Foreword..........1

1.

2.

3.

4.

5.

6.

7.

8.

9.

10.

11.

Foreword

Time has long been one of the greatest mysteries of the universe. While we experience it as a steady flow, modern physics suggests a far more intricate and malleable reality.

From ancient spiritual traditions to the cutting-edge discoveries in quantum mechanics, the nature of time remains a subject of profound exploration.

Throughout my life, I have been intimately acquainted with time in ways that defy traditional explanation. My personal journey-- marked by profound medical experiences and near-death encounters--has provided me with an internal sense of time's fluidity.

This book is a synthesis of scientific inquiry, historical exploration, and philosophical speculation, seeking to bridge the gap between what we **know** and what we **feel**.

Time is not merely an external dimension through which we move; it may be deeply interwoven with **consciousness, perception, and the fundamental structure of reality itself**. This work aims to explore the possibility that time is more than a linear construct--it may be a **wave, a field, or even an emergent property shaped by our interactions with it**.

Join me in uncovering the **hidden nature of time** and the profound implications it holds for our understanding of existence.

Introduction: Understanding the Time Wave Theory

The Evolution of Time Perception

Time has fascinated humanity for millennia, shaping the way we experience existence, measure events, and define reality. Early civilizations viewed time through **cyclical patterns**, often tied to **seasons, lunar cycles, and planetary movements**.

The **Hindu concept of Kala**, the **Mayan Long Count calendar**, and **Greek notions of eternal recurrence** suggested that time was **repetitive rather than linear**.

Ancient Greek philosophers **Heraclitus and Parmenides** were among the first to debate the nature of time.

- **Heraclitus** famously declared, *"No man ever steps in the same river twice,"* implying that time was in **constant flux**.

- **Parmenides**, on the other hand, argued that **change was an illusion**, suggesting a **static and eternal universe**.

These differing viewpoints laid the groundwork for later discussions on **the nature of reality, causality, and determinism**.

The **scientific perception of time** took a major shift with **Isaac Newton** in the 17th century. Newton's classical mechanics presented time as an **absolute, universal constant**, flowing at a steady rate independent of external influences. This **Newtonian view of time** dominated physics for over two centuries, providing a foundational framework for mechanics and astronomy.

However, in the early 20th century, **Albert Einstein revolutionized our understanding of time**. His **theories of Special and General Relativity** demonstrated that time is **not absolute** but rather **relative**--it can stretch, contract, and slow depending on velocity and gravitational fields. This discovery shattered the **Newtonian concept of immutable time**, replacing it with **spacetime, a four-dimensional continuum where time and space are interwoven**.

Time Perception in Ancient Civilizations

While modern science explores the physics of time, early civilizations developed **philosophical and religious interpretations** of time based on their observations of the natural world.

Hindu and Buddhist Cycles of Time

Hindu cosmology presents **Kala (time)** as a **cyclical force**, consisting of **vast repeating cosmic ages**:

- **Satya Yuga (Golden Age)** - An era of enlightenment and virtue.
- **Treta Yuga** - A gradual decline in righteousness.
- **Dvapara Yuga** - Further decay of morality.
- **Kali Yuga** - The present age of darkness and suffering.

Each cycle spans **millions of years**, suggesting a **non-linear, infinite nature of time**.

Buddhist traditions propose that **time is an illusion**, emphasizing **impermanence (Anicca)**--the idea that all things, including time, are constantly in flux.

Mayan Cosmology and Astronomical Cycles

The **Mayan civilization** developed one of the most precise **calendar systems** of the ancient world, recognizing the **cyclical nature of time** through:

- **The Long Count Calendar** - Tracks cycles of **5,125 years**.

- **The Tzolk'in (260-day calendar)** - A spiritual cycle aligned with human consciousness.

- **The Haab' (365-day calendar)** - A solar cycle used for agriculture.

Modern Physics and Time Waves

Building on Einstein's insights, modern physics has continued to challenge our classical notions of time. The emergence of **quantum mechanics, chaos theory, and fractal geometry** suggests that time may not be **strictly linear** but rather **fluid, oscillatory, and multidimensional**. The idea of **time as a wave** has gained traction in various theoretical models.

One of the most intriguing ideas in modern physics is that time may behave **as a fluctuating field rather than a continuous sequence**. In **quantum mechanics**, fundamental principles governing particles suggest that **even time itself may exhibit wave-like behavior**. Could time, much like light, exist as both **a particle and a wave**? If so, what are the implications for **our perception of reality, free will, and the nature of the universe?**

Mathematical Foundations of Time Waves

Mathematical models have begun to describe time as **a fluctuating wave, governed by oscillatory motion and harmonic resonance**. Some key mathematical tools used to explore time waves include:

- **Wave equations** - Fundamental equations in physics describing how **oscillations propagate** through space and time.

- **Fourier transforms** - Used in signal processing and quantum mechanics to break down **complex waveforms into component frequencies**.

- **Fractal geometry** - Explores self-repeating structures that appear at multiple scales, potentially describing time's cyclical nature.

Conclusion: A New Paradigm for Time

The nature of time is one of the **greatest mysteries in physics, philosophy, and human experience**. The time wave theory proposes that time is not **an immutable, linear progression** but rather a **dynamic, oscillating field** that interacts with consciousness, energy, and quantum processes.

If these theories hold, then our understanding of **causality, free will, and reality itself** may need to be re-evaluated. Future research may unlock **new ways of perceiving, manipulating, and even navigating time**.

Chapter 1: Introduction - Understanding the Time Wave Theory

The Evolution of Time Perception

Time has fascinated humanity for millennia, shaping the way we experience existence, measure events, and define reality. Early civilizations viewed time through **cyclical patterns**, often tied to **seasons, lunar cycles, and planetary movements**.

The **Hindu concept of Kala**, the **Mayan Long Count calendar**, and **Greek notions of eternal recurrence** suggested that time was **repetitive rather than linear**.

Ancient Greek philosophers **Heraclitus and Parmenides** were among the first to debate the nature of time.

- **Heraclitus** famously declared, *"No man ever steps in the same river twice,"* implying that time was in **constant flux**.

- **Parmenides**, on the other hand, argued that **change was an illusion**, suggesting a **static and eternal universe**.

These differing viewpoints laid the groundwork for later discussions on **the nature of reality, causality, and determinism**.

The **scientific perception of time** took a major shift with **Isaac Newton** in the 17th century. Newton's classical mechanics presented time as an **absolute, universal constant**, flowing at a steady rate independent of external influences. This **Newtonian view of time** dominated physics for over two centuries, providing a foundational framework for mechanics and astronomy.

However, in the early 20th century, **Albert Einstein revolutionized our understanding of time**. His **theories of Special and General Relativity** demonstrated that time is **not absolute** but rather **relative**--it can stretch, contract, and slow depending on velocity and gravitational fields. This discovery shattered the **Newtonian concept of immutable time**, replacing it with **spacetime, a four-dimensional continuum where time and space are interwoven**.

Time Perception in Ancient Civilizations

While modern science explores the physics of time, early civilizations developed **philosophical and religious interpretations** of time based on their observations of the natural world.

Hindu and Buddhist Cycles of Time

Hindu cosmology presents **Kala (time)** as a **cyclical force**, consisting of **vast repeating cosmic ages**:

- **Satya Yuga (Golden Age)** - An era of enlightenment and virtue.
- **Treta Yuga** - A gradual decline in righteousness.
- **Dvapara Yuga** - Further decay of morality.
- **Kali Yuga** - The present age of darkness and suffering.

Each cycle spans **millions of years**, suggesting a **non-linear, infinite nature of time**.

Buddhist traditions propose that **time is an illusion**, emphasizing **impermanence (Anicca)**--the idea that all things, including time, are constantly in flux.

Mayan Cosmology and Astronomical Cycles

The **Mayan civilization** developed one of the most precise **calendar systems** of the ancient world, recognizing the **cyclical nature of time** through:

- **The Long Count Calendar** - Tracks cycles of **5,125 years**.

- **The Tzolk'in (260-day calendar)** - A spiritual cycle aligned with human consciousness.

- **The Haab' (365-day calendar)** - A solar cycle used for agriculture.

Modern Physics and Time Waves

Building on Einstein's insights, modern physics has continued to challenge our classical notions of time. The emergence of **quantum mechanics, chaos theory, and fractal geometry** suggests that time may not be **strictly linear** but rather **fluid, oscillatory, and multidimensional**. The idea of **time as a wave** has gained traction in various theoretical models.

One of the most intriguing ideas in modern physics is that time may behave **as a fluctuating field rather than a continuous sequence**. In **quantum mechanics**, fundamental principles governing particles suggest that **even time itself may exhibit wave-like behavior**. Could time, much like light, exist as both **a particle and a wave**? If so, what are the implications for **our perception of reality, free will, and the nature of the universe?**

Mathematical Foundations of Time Waves

Mathematical models have begun to describe time as **a fluctuating wave, governed by oscillatory motion and harmonic resonance**. Some key mathematical tools used to explore time waves include:

- **Wave equations** - Fundamental equations in physics describing how **oscillations propagate** through space and time.

- **Fourier transforms** - Used in signal processing and quantum mechanics to break down **complex waveforms into component frequencies**.

- **Fractal geometry** - Explores self-repeating structures that appear at multiple scales, potentially describing time's cyclical nature.

Could Time Waves Explain Anomalous Phenomena?

If time indeed behaves as a **wave**, it could provide a framework for explaining various unexplained phenomena, including:

- **Deja vu** - Could be a result of our consciousness temporarily aligning with another phase of a time wave.

- **The Mandela Effect** - If time waves create **overlapping probabilities**, it may cause groups of people to recall events differently.

- **Premonitions and Precognition** - If time propagates in a wave-like manner, certain individuals might be able to perceive future events through subtle temporal ripples.

- **Time Dilation Beyond Relativity** - Could time waves explain why some individuals report **altered time perception** in deep meditation, near-death experiences, or psychedelic states?

Future Research Directions

The **time wave theory** is still in its infancy, but future research may provide **experimental validation** for these ideas. Some potential areas of study include:

- **Quantum field theory experiments** - Investigating whether time behaves as a quantum field subject to fluctuations.

- **Astrophysical observations** - Studying gravitational waves and their possible connection to time oscillations.

- **Cognitive science and neuroscience** - Examining how brainwave frequencies correlate with subjective time perception.

- **AI and data analysis** - Using machine learning to detect patterns in historical and personal timelines that could hint at cyclical time structures.

Conclusion: A New Paradigm for Time

The nature of time is one of the **greatest mysteries in physics, philosophy, and human experience**. The time wave theory proposes that time is not **an immutable, linear progression** but rather a **dynamic, oscillating field** that interacts with consciousness, energy, and quantum processes.

If these theories hold, then our understanding of **causality, free will, and reality itself** may need to be re-evaluated. Future research may unlock **new ways of perceiving, manipulating, and even navigating time**, leading to advances in **space travel, energy efficiency, and cognitive enhancement**.

The journey to uncover the secrets of time has only just begun.

Chapter 2: The Foundations of Time in Modern Physics

Introduction: Time as a Scientific Concept

For centuries, our understanding of time was shaped by **Newtonian mechanics**, which proposed that time was **absolute, uniform, and independent of space**. According to **Isaac Newton**, time existed as a **background parameter**, flowing at the same rate for all observers, unaffected by motion, gravity, or energy interactions.

Mathematically, Newtonian time is expressed as:

$$t' = t$$

where **t'** represents the time measured by one observer, and **t** represents time measured by another observer. In this classical framework, all observers agree on the passage of time, regardless of their relative motion.

This classical notion of time persisted until the early 20th century when **Albert Einstein introduced Special and General Relativity**. These groundbreaking theories **shattered the idea of absolute time**, demonstrating that time was in fact **relative**--it could be stretched, contracted, and influenced by external forces.

Time Dilation in Special Relativity

Einstein's **Special Theory of Relativity (1905)** introduced the idea that time is **not absolute but depends on the observer's frame of reference**. This led to the concept of **time dilation**, which states that the passage of time **slows down** for an observer moving at speeds close to the speed of light.

The famous **time dilation equation** is given by:

$$Dt' = Dt\, sqrt(1 - v2c2)$$

where:

- **Dt** is the proper time (the time interval measured by a stationary observer).

- **Dt'** is the dilated time (the time interval measured by a moving observer).

- **v** is the velocity of the moving observer.

- **c** is the speed of light.

Experimental Confirmation of Time Dilation

Scientific experiments have confirmed time dilation:

- **Hafele-Keating Experiment (1971)**: Atomic clocks flown on commercial jets experienced measurable time dilation.

- **Particle Accelerator Tests**: High-energy particles, such as muons, decay more slowly when moving at relativistic speeds, consistent with relativistic predictions.

Gravitational Time Dilation in General Relativity

Einstein's **General Theory of Relativity (1915)** extended these ideas by incorporating the effects of **gravity on time**. It showed that time **slows down in stronger gravitational fields**, leading to **gravitational time dilation**. This phenomenon occurs because gravity bends spacetime, affecting how time flows.

The equation for gravitational time dilation is:

$$Dt' = Dt \sqrt{1 - 2GM/rc^2}$$

where:

- **G** is the gravitational constant.
- **M** is the mass of the gravitating body.
- **r** is the radial distance from the center of the mass.
- **c** is the speed of light.

Experimental Confirmation of Gravitational Time Dilation

This has been verified through multiple observations:

- **GPS satellite corrections** - The clocks on GPS satellites run faster than Earth-based clocks due to weaker gravitational influence in orbit.
- **Time measurements near black holes** - Predictions about extreme time dilation near black holes have been supported by astrophysical observations.

Quantum Mechanics and Temporal Superposition

While relativity describes time on a **macroscopic scale**, **quantum mechanics challenges our understanding of time on a microscopic level**. In quantum theory, particles **do not follow definite trajectories**; instead, they exist in **superpositions of multiple states simultaneously**.

The **Schrodinger equation** governs the evolution of quantum systems:

$$i\hbar \, \partial \Psi / \partial t = H\Psi$$

where:

- **Ψ** is the wavefunction describing the quantum state.

- **H** is the Hamiltonian operator (total energy of the system).

- **i** is the imaginary unit.

- **ℏ** is the reduced Planck constant.

One of the most intriguing implications of quantum mechanics is **temporal superposition**--the idea that a particle can exist in multiple **time states simultaneously** until measured. This challenges classical determinism and suggests that time, like position and momentum, may be **probabilistic rather than fixed**.

Implications for the Nature of Reality

The evolution of our understanding of time--from **Newtonian absolutes to relativistic elasticity and quantum indeterminacy**--suggests that time may be more **fluid and interactive** than previously thought. Key implications include:

- **The Illusion of the Present:** If all moments exist in spacetime, is the experience of "now" merely an artifact of consciousness?

- **Time Travel Possibilities:** If closed timelike curves exist, could we one day send information--or even matter--backward or forward in time?

- **Quantum Time Manipulation:** Can quantum mechanics provide mechanisms for influencing time perception or even altering time's flow?

- **Artificial Time Control:** Could future technology allow us to control time the way we manipulate energy today?

Implications for the Nature of Reality

The evolution of our understanding of time—from **Newtonian absolutes to relativistic elasticity and quantum indeterminacy**—suggests that time may be more **fluid and interactive** than previously thought. This leads to profound implications for physics, philosophy, and technology.

Key Implications

• The Illusion of the Present

If all moments exist in spacetime as suggested by the block universe theory, is the experience of "now" merely an artifact of consciousness? Some physicists argue that past, present, and future are equally real, challenging our intuitive sense of a linear flow of time.

• Time Travel Possibilities

If closed timelike curves exist, could we one day send information—or even matter—backward or forward in time? Theoretical frameworks like Gödel's rotating universes and traversable wormholes hint at possibilities that remain speculative but mathematically feasible.

• Quantum Time Manipulation

Can quantum mechanics provide mechanisms for influencing time perception or even altering time's flow? Concepts such as the delayed-choice experiment and retrocausality suggest that the present can influence the past, raising intriguing questions about causality.

• Artificial Time Control

Could future technology allow us to control time the way we manipulate energy today? Speculative technologies like time crystals, temporal cloaking, and quantum field manipulation hint at the potential for direct human intervention in time's natural progression.

Future Directions

Research into these topics is still in its infancy, but advancements in quantum mechanics, general relativity, and emerging computational models may provide deeper insights into time's true nature. If time is not an absolute but a dynamic, emergent property of the universe, our understanding of reality may be on the brink of a paradigm shift.

The Future of Time Research

Introduction

Time has always been one of the most elusive and fascinating subjects in science and philosophy. Our understanding has evolved from a linear progression of moments to a complex interplay of space, energy, and consciousness. Recent breakthroughs in quantum mechanics and cosmology are opening up entirely new ways to think about time.

Conclusion: The Future of Time Research

The study of time remains at the forefront of modern physics. With every discovery, we peel back layers of mystery surrounding its nature. Advances in quantum gravity, high-energy physics, and cognitive science are gradually revealing that time might be far more dynamic than the simple ticking of a clock. These insights are already hinting at practical applications—from revolutionizing space travel to transforming energy production and even altering our understanding of human consciousness.

Emerging Theories in Time Physics

Cutting-edge research suggests that time might not be a basic, standalone fabric of reality but rather an emergent phenomenon. Theories such as loop quantum gravity, the holographic principle, and even aspects of string theory propose that space and time are intertwined in ways that challenge classical views. New models—like causal dynamical triangulation and quantum loop approaches—are providing fresh perspectives, suggesting that the flow of time could arise from the deep, underlying quantum structure of the universe.

Technological Implications of Time Research

As our grasp of time deepens, we may see transformative technological breakthroughs, including:

- **Improved Time Dilation Compensation:** Future spacecraft could employ advanced systems to counteract time dilation effects, making high-speed interstellar travel more feasible without the relativistic pitfalls of conventional physics.
- **Quantum Clocks:** Next-generation clocks based on quantum principles promise unprecedented accuracy, potentially revolutionizing navigation systems, communication networks, and scientific measurements.
- **Manipulation of Causal Structures:** By exploring how causality works at a quantum level, researchers might develop new computing paradigms or secure communication methods that exploit subtle manipulations of cause and effect.
- **Energy Harnessing through Temporal Field Interactions:** Some theories suggest that understanding time's energy dynamics could lead to harnessing new, untapped energy sources, providing radical solutions to global energy challenges.

Philosophical and Cognitive Considerations

The scientific exploration of time also raises profound philosophical and psychological questions. Neuroscience shows that our perception of time is highly subjective, influenced by culture and individual experience. As our models of time become more intricate, they invite us to ask whether we could eventually learn to control our own temporal perception—potentially unlocking new states of awareness and even enhancing cognitive function.

Final Thoughts

The future of time research is not only a scientific frontier but a transformative journey that could redefine our very existence. As breakthroughs converge across physics, technology, and philosophy, we stand on the threshold of a new era—one where time might be something we not only experience but also understand, shape, and perhaps even control.

Chapter 3: Quantum Mechanics and the Illusion of Linear Time

Introduction: The Paradox of Time in Quantum Mechanics

Classical physics describes time as a **continuous, linear flow**, much like an unbreakable arrow moving from the past to the future. However, quantum mechanics presents a radically different perspective--one in which time may not be as rigid or absolute as once believed.

Some of the most perplexing questions in physics involve **the nature of time at the quantum scale**:

- Does time flow smoothly, or is it fundamentally **quantized**?

- Can quantum systems **exist in multiple time states simultaneously**?

- Is the **past influenced by the future**, or is causality an emergent phenomenon?

This chapter explores the ways in which **quantum mechanics challenges our conventional understanding of time**.

The Observer Effect: Does Consciousness Shape Reality?

One of the most controversial aspects of quantum mechanics is the **observer effect**, which suggests that **the mere act of observation can alter the outcome of an event**. This phenomenon challenges our classical view of an objective reality and opens the door to the idea that consciousness may play an integral role in the unfolding of physical events.

At its core, the observer effect raises fundamental questions: Is the universe inherently indeterminate until measured? And if so, does the observer's mind actively participate in collapsing possibilities into a single, definitive outcome?

The Double-Slit Experiment

The **double-slit experiment** is a landmark demonstration in quantum mechanics. In this experiment, particles such as photons or electrons are fired toward a barrier with two parallel slits. When both slits are open and no measurement is made, the particles create an interference pattern on a detection screen—a signature of wave-like behavior. However, when an observer or measuring device is introduced to determine which slit a particle passes through, the interference pattern vanishes, and the particles appear to follow classical, particle-like trajectories.

This dramatic change in behavior, contingent solely on observation, implies that the act of measurement plays a non-trivial role in the manifestation of reality.

Mathematically, the behavior of particles in their wave-like state can be expressed by the wavefunction:

$$\Psi(x,t) = A\, e^{i(kx - \omega t)}$$

where:

- $\Psi(x,t)$ represents the probability amplitude of the particle.
- **A** is the amplitude of the wave.
- **k** is the wave vector, which is related to the momentum of the particle.
- ω (omega) is the angular frequency, connected to the energy of the particle.

When an observation is made, the wavefunction **collapses** into a single state, implying that the act of measurement determines which outcome is realized. This collapse is at the heart of debates about the role of consciousness in quantum mechanics.

Various interpretations attempt to explain this behavior. The *Copenhagen interpretation* posits that the observer causes the collapse of the wavefunction, while the *many-worlds interpretation* suggests that all possible outcomes occur in separate branches of reality, eliminating the need for a conscious observer to cause collapse.

These interpretations, while differing in mechanism, both underscore the mysterious interplay between observation and physical reality. They challenge the conventional separation between the observer and the observed, hinting that consciousness might be more deeply woven into the fabric of the universe than previously thought.

In light of these findings, the profound question arises: *Does consciousness actively shape the physical world, or is it simply a passive bystander to processes that unfold independently of our awareness?* Ongoing research in quantum mechanics, neuroscience, and philosophy continues to probe this question, striving to uncover the true nature of reality.

Quantum Superposition and Temporal Ambiguity

One of the defining principles of quantum mechanics is **superposition**, the idea that a particle can exist in **multiple states simultaneously** until measured.

Schrodinger's Cat: A Thought Experiment in Temporal Superposition

Physicist **Erwin Schrodinger** proposed a famous thought experiment involving a cat placed in a sealed box with a radioactive atom that has a 50% chance of decaying within an hour.

- If the atom decays, the cat dies.
- If the atom does not decay, the cat remains alive.

According to quantum mechanics, until observed, the cat exists in a **superposition of both alive and dead states**. When an observer checks inside the box, the wavefunction collapses, and the cat is found in one definite state.

Time as a Probabilistic Construct

If quantum superposition applies to time, **could past, present, and future states coexist** until observed? Some theories suggest that:

- The **future may influence the past** in ways we do not yet fully understand.
- Time is **nonlinear** and can exist in multiple probabilities simultaneously.

The Delayed-Choice Quantum Eraser: Retrocausality in Action

Another experiment that challenges our understanding of time is the **Delayed-Choice Quantum Eraser**.

How It Works

- A particle is emitted toward a **beam splitter**, which directs it into one of two paths.

- Before reaching the detector, the experimenter has the choice to measure which path the particle took.

- If measured, the particle behaves like a particle. If not, it behaves like a wave.

The Mind-Bending Implication

If the experimenter chooses to measure the path **after the particle has already traveled**, the wavefunction still collapses--**as if the past had been altered retroactively**.

This suggests that **the future can influence the past**, breaking the conventional laws of causality.

Quantum Entanglement and Time Symmetry

The "Spooky Action at a Distance" Phenomenon

Quantum entanglement occurs when two particles become **instantaneously correlated**, regardless of the distance between them. If one entangled particle's state is determined, the other's state **is immediately known**, even if separated by vast cosmic distances.

Einstein famously referred to this as **"spooky action at a distance**," believing it contradicted the classical understanding of time.

Implications for Time

- If quantum entanglement occurs **instantaneously**, does this mean **time is irrelevant** at the quantum level?

- Could entanglement be evidence of a **timeless reality**, where past, present, and future are interwoven?

Some physicists propose that **quantum entanglement may extend across time as well as space**, allowing **future events to influence past outcomes**.

Chapter 3: Quantum Mechanics and the Illusion of Linear Time

Introduction: The Paradox of Time in Quantum Mechanics

Classical physics describes time as a **continuous, linear flow**, much like an unbreakable arrow moving from the past to the future. However, quantum mechanics presents a radically different perspective—one in which time may not be as rigid or absolute as once believed.

Some of the most perplexing questions in physics involve **the nature of time at the quantum scale**:

- Does time flow smoothly, or is it fundamentally **quantized**?

- Can quantum systems **exist in multiple time states simultaneously**?

- Is the **past influenced by the future**, or is causality an emergent phenomenon?

The Many-Worlds Interpretation: Does Time Branch Infinitely?

Hugh Everett's Radical Proposal

The **Many-Worlds Interpretation (MWI)** suggests that every quantum event **splits the universe into multiple parallel timelines**, each representing a different outcome.

> Instead of wavefunction collapse, all possibilities play out in separate, branching realities.

For example:

- In one universe, you choose **coffee** this morning.
- In another, you choose **tea**.
- Both realities exist independently, but you only experience one.

If MWI is correct, then **time is not a singular path but an infinitely branching structure**.

Does Every Decision Create a New Timeline?

If every decision creates a new timeline, then:

- **Time does not move forward; it expands.**
- **Our consciousness follows a singular thread through an ever-diverging multiverse.**

This raises profound questions:

- Can we influence which timeline we follow?
- Could we access or traverse other timelines?

Implications for Reality and Technology

If time is indeed more fluid than traditionally thought, future technologies might exploit quantum time manipulation. Concepts such as quantum computing, temporal cloaking, and even time crystals hint at practical applications that may one day redefine our experience of time.

As research continues to unfold, our understanding of time may be on the brink of a revolutionary shift, bridging the gap between classical determinism and quantum uncertainty.

Time Crystals: Evidence for Temporal Quantization?

The Discovery of Time Crystals

In 2012, physicist **Frank Wilczek** proposed the existence of **time crystals** – a novel state of matter that oscillates in time without the need for energy input. Unlike conventional crystals, which display a repeating pattern in space, time crystals exhibit periodic motion in time.

This groundbreaking idea challenges the traditional view of time as a continuous flow, instead suggesting that time may be quantized. In other words, just as matter exists in discrete packets (atoms and molecules), time itself might be composed of individual, discrete moments or "ticks" that repeat in a regular, clockwork-like fashion.

Experimental realizations have since been demonstrated in controlled quantum systems, such as trapped ions and spin systems, further fueling debates and research into the fundamental nature of time.

Implications for Time Waves

If time crystals exist, they could provide evidence that:

- **Time may be quantized** – suggesting that, like energy, time is not continuous but consists of discrete intervals.

- **Certain phenomena may repeat periodically** – much like standing waves, where recurring patterns emerge over time without external driving forces.

Such observations could indicate that the fabric of time has a wave-like structure, fundamentally altering our understanding of temporal dynamics.

Practical Implications and Future Prospects

The potential applications of time crystals extend far beyond theoretical physics:

- **Quantum Computing:** Time crystals may function as perfect oscillators, providing ultra-stable time references that are crucial for synchronizing quantum bits and reducing decoherence.

- **Time Travel Theories:** The notion of discrete time intervals may offer insights into cyclic or reversible processes, potentially informing speculative models of time travel or retrocausality.

- **New Physics:** Uncovering a quantized nature of time could pave the way for breakthroughs in energy harvesting, novel materials, and our overall understanding of the universe's fundamental laws.

As research continues, time crystals might not only validate the concept of temporal quantization but also inspire entirely new technologies and scientific paradigms.

Conclusion: The Quantum Nature of Time

Quantum mechanics challenges traditional notions of **causality, determinism, and the linear progression of time**. Unlike classical physics, which treats time as a steady, unidirectional flow, quantum physics suggests that time is **not an independent, fixed quantity**, but rather a **flexible, interactive component of reality**.

At the heart of this shift is the realization that time, space, and even reality itself may be **observer-dependent**. The quantum world defies intuition, allowing for the possibility that **multiple timelines, retrocausality, and even timeless states** may coexist within a broader framework of existence.

Key Takeaways

- **The observer effect** implies that **reality, including time, may be dependent on observation**. This raises the question: *Does time exist independently, or does it require an observer?*

- **Quantum superposition suggests** that multiple time states may exist simultaneously. If true, this could mean that the experience of a single "now" is an illusion.

- **The Delayed-Choice Quantum Eraser experiment** hints that the **future may influence the past**. This suggests that cause and effect may be more fluid than we assume.

- **Quantum entanglement defies conventional time constraints**, hinting at **instantaneous communication across space and time**. If information can be shared beyond the limits of relativity, is time truly a limiting factor?

- **The Many-Worlds Interpretation** suggests that time may be **an infinitely branching structure rather than a single timeline**. This raises fascinating implications about fate, free will, and the nature of reality.

- **Time crystals suggest** that time may be **quantized rather than continuous**. If time itself has a discrete structure, our perception of smooth temporal flow may be an emergent phenomenon.

Emerging Theories and the Future of Time Research

The nature of time is one of the greatest unsolved mysteries in physics. Scientists continue to explore revolutionary ideas that may reshape our understanding of time and its role in the universe.

Holographic Time and the End of Temporal Duality

The **holographic principle** suggests that our perception of time is merely a projection of a deeper, more fundamental reality. If this is correct, then time itself may be **a byproduct of quantum information**, rather than an intrinsic property of the universe.

In some models, time emerges from entangled quantum states, much like how spatial dimensions emerge in string theory. This implies that **the past, present, and future may be encoded in a higher-dimensional structure**, similar to a hologram.

The Role of Consciousness in Time Perception

An intriguing question arises: *Is time something that exists independently, or is it constructed by the mind?*

Some physicists propose that time is a function of **consciousness and memory**. Without memory, would time exist at all? Experiments in neuroscience show that the brain may actively construct its sense of time by integrating sensory data and predicting future states.

Could this mean that **time is not fundamental but an emergent phenomenon linked to cognition**? If so, understanding consciousness may be the key to understanding time itself.

Final Thoughts and Philosophical Implications

The study of time has always been at the intersection of **physics, philosophy, and metaphysics**. Quantum mechanics has shattered many classical assumptions, leaving us with profound questions:

- If the future can influence the past, what does this mean for free will?

- If time is an emergent property, could we one day manipulate it?

- If time is an illusion, what is the ultimate nature of reality?

Modern physics is on the verge of unlocking some of these mysteries. As we continue exploring the **frontiers of quantum mechanics, consciousness, and time physics**, we may one day discover that **time is not what we once believed it to be**.

What remains certain is that **the relationship between time, space, and reality is far more complex than we ever imagined**. Our journey into understanding time is not only a scientific endeavor—it is a journey into the very fabric of existence itself.

Chapter 4: Free Energy and Nikola Tesla's Vision

Introduction: The Quest for Limitless Energy

Human civilization has long been **dependent on finite energy resources** such as fossil fuels, nuclear power, and conventional electricity grids. But what if unlimited, freely available energy were possible? Could we harness **the natural forces of the universe** to create a **self-sustaining, wireless energy system** that would transform every aspect of modern life?

The dream of tapping into an endless energy source not only promises environmental and economic benefits but also the possibility of reshaping geopolitical power structures and liberating societies from energy scarcity.

Tesla's Vision and Innovative Research

One of the most revolutionary thinkers in this area was **Nikola Tesla**. More than a prolific inventor, Tesla was a visionary whose pioneering work in electricity and magnetism challenged established ideas and opened new possibilities. His ambitious projects sought to make energy:

- **Wireless and globally distributed** – envisioning a network where energy flows through the atmosphere, bypassing the need for traditional cables and power stations.

- **Extracted from the natural electromagnetic field of the Earth** – suggesting that our planet's inherent energy, if properly harnessed, could be a continuous and renewable resource.

- **Potentially limitless and freely available to all humanity** – offering a future where energy scarcity could become a relic of the past.

Zero-Point Energy and the Quantum Realm

Central to the discussion of free energy is the concept of **zero-point energy** – the idea that even in a vacuum, a fluctuating energy field exists due to quantum mechanics. While extracting usable energy from this phenomenon remains a theoretical challenge, it suggests that the very fabric of space might harbor an immense, untapped resource.

Researchers continue to debate whether advancements in quantum physics could eventually lead to practical methods for harnessing this energy, potentially revolutionizing the energy sector in ways that Tesla only dreamed about.

Scientific and Political Barriers to Free Energy

Despite the allure of limitless energy, significant obstacles remain. On the scientific front, the extraction and conversion of energy from ambient electromagnetic fields or the quantum vacuum involve challenges that stretch the limits of current technology and understanding.

Equally daunting are the political and economic challenges. Established energy industries, regulatory frameworks, and vested interests can all pose formidable barriers to the adoption of revolutionary energy systems. The transition from a fossil fuel-based economy to one powered by free energy would require not only technological breakthroughs but also bold shifts in policy and global cooperation.

Tesla's Legacy and Modern Implications

Today, Tesla's work continues to inspire scientists, engineers, and futurists. His ideas have transcended the boundaries of his era, prompting ongoing exploration into sustainable energy, wireless power transmission, and the possibilities offered by the quantum world.

Modern initiatives in renewable energy and advancements in wireless technology are, in many ways, the heirs to Tesla's dream. Whether through solar power innovations, advancements in battery technology, or experimental projects in wireless energy transmission, the legacy of Tesla's visionary ideas is evident in the persistent pursuit of a cleaner, more equitable energy future.

Conclusion

The exploration of free energy is as much about challenging our scientific paradigms as it is about reimagining our socio-economic structures. As research progresses and the boundaries of what is possible continue to expand, the vision of harnessing limitless energy may one day emerge from the realm of science fiction into everyday reality. Tesla's legacy endures as a beacon for all those who dare to envision a world powered by the infinite energy of nature.

Tesla's Dream of Wireless Energy Transmission

The Wardenclyffe Tower: A Suppressed Breakthrough?

Nikola Tesla's most ambitious energy experiment was the **Wardenclyffe Tower**, a **1901 project** designed to demonstrate that electrical energy could be **transmitted wirelessly over long distances**.

Tesla believed that by **tapping into the Earth's ionosphere**, energy could be:

- **Sent through the air, eliminating power lines**.
- **Used globally without energy loss**.
- **Freely available to anyone, anywhere on the planet**.

However, the project was **shut down and dismantled**, largely due to **financial and political opposition from industrial energy monopolies**.

The Tesla Coil and High-Frequency Energy

One of Tesla's most famous inventions, the **Tesla Coil**, was an experimental device that could generate and transmit **high-voltage, high-frequency energy** through the air.

Equation for Tesla Coil Resonance:

$f = 1 / (2p[x](LC))$

where:

- **f** is the resonant frequency,

- **L** is the inductance of the coil,
- **C** is the capacitance.

The Tesla Coil demonstrated the **fundamental principle of resonant energy transfer**, which later became the foundation for **modern wireless charging technologies**.

Zero-Point Energy: A Potential Infinite Power Source

What is Zero-Point Energy?

Quantum mechanics suggests that even "empty space" is **not truly empty** but filled with **fluctuating energy fields**. This is known as **Zero-Point Energy (ZPE)**.

The energy density of the quantum vacuum is given by:

$$r\_vacuum = hc / (2p)3 \ [x]0^\wedge[x] \ k3 \ dk$$

where:

- **h** is the reduced Planck's constant,
- **c** is the speed of light,
- **k** is the wave vector.

If harnessed, zero-point energy could provide a **perpetual, limitless power source**.

The Casimir Effect: Evidence of Vacuum Energy

The **Casimir Effect** demonstrates that quantum fluctuations create **measurable forces**, even in a vacuum. Two closely spaced, uncharged metal plates experience **an attractive force due to quantum vacuum fluctuations**.

Could this force be **scaled up and harnessed** for **practical energy production**?

The Challenges and Controversies Surrounding Free Energy

Despite Tesla's discoveries and modern quantum energy theories, **free energy research faces significant challenges**.

Scientific Skepticism and Theoretical Barriers

Many physicists argue that extracting free energy from the vacuum **would violate the laws of thermodynamics**. Critics claim that:

- **Zero-point energy is real**, but extracting usable energy from it may be impossible.

- Wireless energy **loses efficiency over large distances**.

- Large-scale energy extraction **could disrupt natural electromagnetic fields**.

Industrial and Political Opposition

A major obstacle to free energy has been **corporate and governmental control over the energy industry**. Historically:

- **Wardenclyffe Tower was defunded** because investors, such as **J.P. Morgan**, could not monetize free electricity.

- **Patent suppression and corporate acquisitions** have prevented certain energy technologies from reaching public markets.

- **Fear of economic disruption** has led to **classified energy research**, limiting public access to alternative energy discoveries.

Alternative Free Energy Theories and Technologies

Apart from Tesla's work and zero-point energy, other **alternative energy concepts** have emerged.

Cold Fusion: The Controversial Energy Breakthrough

Cold fusion, or **Low-Energy Nuclear Reaction (LENR)**, proposes that nuclear reactions can occur at room temperature, **releasing enormous amounts of energy without harmful radiation**.

- The **Pons and Fleischmann Experiment (1989)** initially suggested cold fusion was possible.

- Mainstream physicists **dismissed the results**, but independent researchers have since reported **excess heat production**.

Could cold fusion one day provide a **safe, limitless energy source**?

Torsion Field Energy and Scalar Waves

Some researchers propose that **torsion fields**--hypothetical spiral energy structures--could be a previously unknown form of **subtle energy**.

- Scalar waves, first described by **James Clerk Maxwell**, could represent an **alternative type of electromagnetic radiation** capable of **transmitting energy without loss**.

While these ideas remain speculative, continued research may reveal **undiscovered forces in the physics of energy transmission**.

Future Research and Applications of Free Energy

The next steps in **free energy research** should focus on:

1. **Validating zero-point energy extraction experiments**.
2. **Developing high-efficiency wireless energy transfer technologies**.
3. **Exploring alternative energy fields beyond classical physics**.
4. **Overcoming political and industrial resistance** to free energy technologies.

How Free Energy Could Transform Society

If free energy were fully developed and released, it would **revolutionize civilization** by:

- **Eliminating fossil fuel dependency**.
- **Providing unlimited, clean energy for all nations**.
- **Accelerating space travel** by enabling **electromagnetic propulsion systems**.
- **Allowing decentralized energy access**, empowering communities worldwide.

Conclusion: A Future of Unlimited Energy?

Tesla's vision of **a world powered by free energy** remains one of the most compelling technological dreams in history. Whether through **zero-point energy, atmospheric electricity, or advanced quantum field manipulation**, the search for **limitless, sustainable energy** continues.

As future scientists push the boundaries of what is possible, we may one day achieve Tesla's ultimate goal: **An energy system that is wireless, free, and available to all.**

Key Takeaways:

- **Tesla's wireless energy experiments** suggested the Earth could transmit power globally.

- **Zero-point energy is real**, but extracting it in usable form remains an open question.

- **The Casimir Effect confirms quantum vacuum energy fluctuations**.

- **Cold fusion and torsion fields may provide new energy solutions**.

- **Industrial and governmental interests have historically suppressed free energy research**.

The next century may hold breakthroughs that could **free humanity from energy scarcity** and usher in a new era of **technological abundance**.

Chapter 5: Warp Bubbles and the Alcubierre Drive

Introduction: The Challenge of Faster-Than-Light Travel

Interstellar travel remains one of the greatest challenges facing humanity. According to **Einstein's theory of relativity**, no object with mass can reach or exceed the speed of light without requiring **infinite energy**. This presents a fundamental barrier to **exploring distant star systems**.

However, theoretical physics suggests that **space itself can be manipulated** to allow apparent **faster-than-light (FTL) travel** without violating relativity. The most well-known of these models is the **Alcubierre Drive**, which involves creating a **warp bubble**--a region of space that contracts in front of a spacecraft and expands behind it, effectively **moving space itself**.

This chapter explores:

- **The science of warp bubbles and the Alcubierre Drive.**
- **The energy challenges associated with FTL travel.**
- **Potential experimental approaches to spacetime engineering.**

The Alcubierre Metric: A Warp Drive Solution

In 1994, physicist **Miguel Alcubierre** proposed a solution to Einstein's field equations that allows for apparent **faster-than-light motion** while remaining consistent with General Relativity.

The Alcubierre metric is given by:

$$ds^2 = -c^2 dt^2 + (dx - v_s f(r_s) dt)^2 + dy^2 + dz^2$$

where:

- **ds2** is the spacetime interval.
- **v_s** is the velocity of the warp bubble.
- **f(r_s)** is a function defining the shape of the bubble.
- **dx, dy, dz** are spatial coordinates.

This metric describes a **localized distortion of spacetime**, where space is **compressed ahead of the spacecraft and expanded behind it**. The spacecraft remains within a **flat region of spacetime**, meaning it experiences **no acceleration** and avoids the effects of time dilation.

Implications of the Alcubierre Drive

- The spacecraft would **not technically move faster than light**, but space itself would **move around it**.
- A traveler within the warp bubble would feel **no acceleration**, making high-speed travel feasible.
- **No violation of relativity** occurs because the spacecraft remains within a locally inertial frame.

Energy Requirements and the Role of Exotic Matter

One of the major challenges of the Alcubierre Drive is that it **requires exotic matter with negative energy density** to function.

The energy required to generate a warp bubble is given by:

$$T_{\mu\nu} u^\mu u^\nu < 0$$

where:

- $T_{\mu\nu}$ is the stress-energy tensor.
- u^μ is the four-velocity of the observer.

This inequality implies that **negative energy density** is necessary--something not yet found in macroscopic quantities.

The Casimir Effect and Negative Energy

One possible source of negative energy is the **Casimir effect**, a quantum phenomenon where two uncharged metal plates placed close together experience an attractive force due to vacuum fluctuations.

The energy density between the plates is given by:

$$\rho_{Casimir} = -\hbar c \pi^2 / (240 d^4)$$

where:

- $\hbar$ is the reduced Planck's constant.
- c is the speed of light.
- d is the distance between the plates.

While the Casimir effect demonstrates that negative energy exists, **scaling this up to the levels required for a warp drive remains a significant challenge**.

Time Travel Implications of Warp Bubbles

One of the most intriguing consequences of a warp bubble is its potential to **manipulate time**. If space can be bent for FTL travel, could similar distortions create **closed timelike curves (CTCs)**, allowing for **time travel**?

Causality Violations and the Grandfather Paradox

If an Alcubierre Drive could be used for interstellar travel, some models suggest that a **return trip could place the traveler in the past**. This introduces potential paradoxes:

- **Grandfather Paradox** - What happens if a traveler arrives in the past and alters history?

- **Predestination Paradox** - Does a time traveler become locked into events that must occur?

Hawking's Chronology Protection Conjecture

Physicist **Stephen Hawking** proposed the **Chronology Protection Conjecture**, which suggests that quantum effects might prevent CTCs from forming, thereby preserving causality. However, this remains speculative.

Current Research and Challenges

Several research teams are actively investigating the feasibility of warp bubbles. Key challenges include:

- **Energy Requirements** - Initial estimates suggested that a warp drive would require energy **greater than the mass of the observable universe**. However, newer models propose that **optimized bubble shapes** could dramatically reduce energy needs.

- **Stability Issues** - The stability of the warp bubble remains uncertain. Some models suggest that quantum fluctuations could cause it to collapse.

- **Negative Energy Sourcing** - While the Casimir effect proves negative energy exists, no known method currently allows for the **large-scale production** of negative energy needed for warp travel.

NASA's Eagleworks Lab and Recent Breakthroughs

NASA's **Eagleworks Laboratory**, led by physicist **Harold White**, has been investigating **warp field mechanics**. In 2011, White proposed modifications to Alcubierre's metric that could potentially reduce energy requirements by orders of magnitude.

Recent advancements include:

- **Lab-scale experiments on spacetime distortions**.

- **New theoretical models suggesting that ring-shaped energy distributions** could create more efficient warp fields.

- **Discovery of stable energy configurations** that might reduce the reliance on exotic matter.

Implications for the Future of Space Travel

The development of **faster-than-light (FTL) travel** would redefine humanity's relationship with the cosmos. While currently hypothetical, technologies such as the **Alcubierre Drive** suggest that manipulating spacetime itself could make interstellar travel practical. The successful engineering of a warp drive would enable **unprecedented exploration, colonization, and transformation of human civilization**.

If a functional Alcubierre Drive can be developed, it would revolutionize space travel by enabling:

- **Interstellar Exploration** - Humanity could reach distant star systems in **days or weeks** rather than thousands of years.

- **Instantaneous Point-to-Point Travel** - If warp bubbles can be stabilized, they could allow for near-instantaneous travel across vast distances.

- **New Understandings of Physics** - Studying warp fields could lead to deeper insights into the fundamental nature of spacetime and energy.

Engineering Challenges and Current Developments

While the **mathematics of warp drives** suggest that FTL travel is possible under general relativity, major **engineering and energy** challenges must be overcome:

- **Exotic Matter and Negative Energy** - A functional warp drive requires **negative energy density**, which may not naturally exist in the required quantities.

- **Energy Requirements** - Early calculations suggested that an Alcubierre Drive would need **more energy than is available in the entire observable universe**. However, recent refinements suggest that lower energy designs could be viable.

- **Structural Stability** - Maintaining the shape of a warp bubble without it collapsing or harming the spacecraft within is an open problem.

Current research at organizations such as NASA's **Advanced Propulsion Physics Laboratory (Eagleworks)** and independent theoretical physicists are exploring ways to **reduce energy demands** and find practical approaches to bending spacetime.

Conclusion: Are Warp Bubbles Science Fiction or Future Reality?

While the Alcubierre Drive remains a **theoretical concept**, ongoing research into **negative energy, quantum field manipulation, and spacetime engineering** suggests that **warp bubbles might one day become a scientific reality**. This burgeoning field of study has ignited interest across both academic and industrial sectors, inspiring researchers to explore the frontiers of faster-than-light travel.

Theoretical Foundations and Scientific Inquiry

The concept of warp bubbles is rooted in Einstein's theory of general relativity, which allows for the deformation of spacetime. The Alcubierre Drive, in particular, envisions a mechanism whereby space in front of a spacecraft is contracted while space behind it is expanded. This creates a bubble of flat spacetime in which the vessel is carried along, potentially allowing for effective faster-than-light travel without violating the laws of relativity.

Central to this theory is the role of **negative energy**—a hypothetical form of energy that would counteract gravitational forces and enable the manipulation of spacetime on a cosmic scale.

Engineering Challenges and Technological Hurdles

Translating the elegant mathematics of the Alcubierre Drive into practical technology faces enormous obstacles. The generation and control of the required **negative energy** pose significant engineering challenges. Current experimental research in quantum field theory and advanced materials science aims to uncover potential methods to harness such exotic energy forms.

Moreover, sustaining a stable warp bubble over interstellar distances would demand innovations in control systems and energy management that far exceed our present capabilities.

Potential Implications for Space Exploration

If warp drive technology is realized, it could revolutionize space travel by eliminating the constraints of conventional propulsion systems. Beyond reducing travel times between distant celestial bodies, the ability to manipulate spacetime could even open avenues for breakthroughs in time manipulation and communication across vast interstellar distances.

Such advances would not only transform our approach to exploration but might also lead to profound shifts in technology and global cooperation, as humanity steps into a new era of space exploration.

Key Takeaways:

- **The Alcubierre Drive proposes a method for FTL travel without violating relativity.**
- **It relies on the manipulation of spacetime itself rather than traditional propulsion.**
- **The biggest challenge remains energy requirements, particularly the need for negative energy.**
- **If developed, warp drives could eliminate travel time constraints and even lead to breakthroughs in time manipulation.**

As our understanding of **quantum mechanics and spacetime continues to evolve**, what was once dismissed as pure science fiction may soon serve as the foundation for revolutionary advances in space exploration.

Chapter 6: Simulation Theory and the Nature of Reality

Introduction: Are We Living in a Simulation?

One of the most provocative questions in modern science and philosophy is whether our reality is **an artificial construct, created by an advanced intelligence or higher-dimensional beings**. With the rise of **artificial intelligence, virtual reality, and quantum physics**, the idea that reality itself is **a highly sophisticated simulation** has gained increasing traction.

This chapter explores:

- **The philosophical and historical origins of simulation theory.**

- **Scientific evidence suggesting reality behaves like a simulation.**

- **Quantum mechanics, digital physics, and the structure of reality.**

The Philosophical Foundations of Simulation Theory

The idea that **reality is an illusion or artificial construct** dates back to ancient philosophy.

Plato's Allegory of the Cave

In **Plato's Republic**, the philosopher describes a group of people who live chained in a cave, only seeing **shadows projected onto a wall**. These shadows represent their entire reality, while **the true world exists beyond their perception**.

This allegory serves as an early form of Simulation Theory:

- **The shadows** = The limited reality we experience.

- **The external world** = The "true" reality, hidden from us.

- **The prisoners** = Us, bound by our perceptions and cognitive limitations.

Descartes' Evil Demon and the Illusion of Reality

In the 17th century, **Rene Descartes** asked: "How do we know reality is real?" He proposed the thought experiment of an **Evil Demon**, an all-powerful being that deceives us into believing in a false world.

His argument suggests that:

- **Sensory experience is unreliable**--we cannot trust our perceptions alone.

- **External reality may be an illusion**--a creation of an external intelligence.

Scientific Evidence for a Simulated Universe

While ancient philosophy hinted at the illusion of reality, modern science has provided **mathematical and computational arguments** supporting the idea that **the universe behaves like a simulation**.

Nick Bostrom's Simulation Hypothesis

Oxford philosopher **Nick Bostrom** formalized the Simulation Argument in 2003, proposing the following trilemma:

1. Civilizations **never reach a post-human stage** capable of running simulations.

2. Post-human civilizations exist but **choose not to run ancestor simulations**.

3. We are almost certainly **living in a simulation** because simulated realities vastly outnumber base reality.

Given that computing power continues to increase exponentially, Bostrom argues that **advanced civilizations could create perfect simulations of their ancestors**, meaning the probability of us being in the "original" reality is extremely low.

Quantum Mechanics and the Pixelated Universe

Quantum mechanics provides some of the strongest evidence that reality may be **computational in nature**.

Planck Scale and the Discrete Nature of Reality

The smallest measurable unit of space is the **Planck length** (*1.6 x 10[x]35 meters*). This suggests that space is **quantized**, much like pixels on a screen.

Equation for Planck Length:

$L_p = [x](hG/c^3)$

where:

- **h** is the reduced Planck constant.
- **G** is the gravitational constant.
- **c** is the speed of light.

If space is not continuous but consists of discrete units, this suggests **a computational limit**, as if reality operates on a grid of fundamental "pixels."

The Double-Slit Experiment and Observer Effect

The famous **double-slit experiment** shows that particles behave **like waves until observed**, suggesting that reality only resolves into a definite state **when measured**.

Wavefunction Representation:

$Ps(x,t) = e^{\wedge}(i(kx - ot))$

where:

- **Ps(x,t)** represents the probability wave of a particle.
- **k** is the wave vector.
- **o** is the angular frequency.

This experiment suggests that **reality does not exist in a definite form until measured**, much like how objects in a video game only render when observed.

Cosmic Fine-Tuning and Mathematical Universality

Many physicists have noted that the **constants of nature**--such as the gravitational constant and the fine-structure constant--are **precisely tuned** for the existence of life.

If gravity were slightly weaker, galaxies would never form.

If the nuclear strong force were slightly different, atoms would not exist.

This fine-tuning raises the question: **Was our universe designed?** A simulated reality would explain why the universe appears **"programmed" with exact physical laws**.

Why Would an Advanced Civilization Create a Simulation?

If we are living in a simulation, the next question is: **Why was it created?** Possible explanations include:

- **Ancestor Simulations** - Future civilizations could be running simulations to study their history.

- **Artificial Consciousness Experiments** - The simulation may be testing how intelligence evolves.

- **Entertainment or Research** - The simulation might be part of an advanced civilization's computational project.

- **A Training Program** - Similar to philosophical and religious ideas, our reality may be a test for higher forms of existence.

The Simulation's Architecture: What Controls Reality?

If the universe is simulated, it would require:

1. **A computational framework** for processing reality.

2. **A physics engine** governing energy, matter, and causality.

3. **Observers or "players"** interacting with the system.

Evidence for Computational Reality

- **Planck scale granularity suggests discrete space-time units.**

- **Quantum entanglement may be a data compression mechanism.**

- **The observer effect implies reality is rendered only when needed.**

Could We Hack the Simulation?

If reality is programmable, could we:

- **Alter fundamental laws of physics?**

- **Interact with the simulation's creators?**

- **Access other simulated realities?**

Some researchers suggest that **neural interfacing, lucid dreaming, and consciousness expansion** could be ways to **interact with the underlying code of reality**.

Conclusion: What If We Discover We Are in a Simulation?

If we one day confirm that reality is simulated, it would raise profound implications:

- **Could we communicate with the simulator?**
- **Can we hack reality and alter its parameters?**
- **Does this mean there is an external "real world" beyond our simulation?**

Key Takeaways:

- **Simulation Theory suggests we may be living in a computationally generated reality**.
- **Quantum physics supports the idea that reality is digital rather than continuous**.
- **If the universe is simulated, it raises questions about who created it and why**.

As computing advances and our understanding of reality deepens, we may one day uncover the true nature of our existence--whether as physical beings or digital constructs in a cosmic experiment.

Chapter 7: The Mandela Effect and Evidence of Temporal Shifts

Introduction: What is the Mandela Effect?

The **Mandela Effect** refers to a phenomenon where **large groups of people remember historical events, names, or details differently from recorded history**. The term was coined by **Fiona Broome**, who discovered that many people **incorrectly remembered** Nelson Mandela dying in prison in the 1980s, despite his **actual** death occurring in 2013.

This phenomenon raises several intriguing questions:

- **Are these false memories a result of cognitive distortions?**

- **Could they be evidence of parallel realities merging?**

- **Is there a deeper connection between collective consciousness and shifting timelines?**

This chapter explores the **scientific, psychological, and quantum-based explanations** for the Mandela Effect and its possible connection to **temporal shifts and altered realities**.

Examples of the Mandela Effect

Some of the most famous examples of the Mandela Effect include:

Historical Events and Celebrity Deaths

Many people recall **different versions of history**, such as:

- **Nelson Mandela's Death** - Some recall him dying in the 1980s, even remembering news reports of a funeral.

- **The Moon Landing Date** - Some claim it happened in 1969, while others swear it was later.

- **Celebrity Deaths** - Some people strongly recall public figures dying at a different time than the official record states.

Changes in Logos and Branding

- **The Berenstain Bears** vs. **The Berenstein Bears** - Many remember the spelling as "Berenstein" rather than "Berenstain."

- **Febreze or Febreeze?** - Some recall an extra "e" in the name of the air freshener brand.

- **Looney Tunes or Looney Toons?** - A common misconception where people remember "Toons" instead of "Tunes."

Movie Quotes and Pop Culture

- **Star Wars:** "Luke, I am your father." - The actual quote is "No, I am your father."

- **Snow White:** "Mirror, mirror on the wall..." - The actual quote is "Magic mirror on the wall..."

- **Monopoly Man:** Many recall him having a monocle, but the official character never did.

The widespread nature of these alternate memories suggests that **either reality has changed, or our perception of it is flawed**.

Scientific Explanations: The Psychology of False Memories

While the Mandela Effect is often attributed to **paranormal or metaphysical causes**, psychologists suggest **cognitive errors** as a possible explanation.

Confabulation and Memory Reconstruction

Memory is **not a perfect recording** but an **active process of reconstruction**. Each time we recall an event, we unknowingly modify details, sometimes integrating **external influences, assumptions, and biases**.

Collective False Memory Formation

Large groups of people can develop **false shared memories** due to:

- **The Misinformation Effect** - Exposure to incorrect information alters recall.

- **Group Reinforcement** - When enough people believe a false fact, others begin accepting it as truth.

- **Cognitive Biases** - Our brains tend to seek patterns, filling in missing details with assumptions.

While these psychological factors explain some aspects of the Mandela Effect, they do not account for **cases where physical records and official documentation appear to have changed**.

Quantum Mechanics and Temporal Shifts

Quantum physics provides an alternative explanation for the Mandela Effect--**that multiple realities may be interacting**.

Quantum Superposition and the Many-Worlds Hypothesis

In quantum mechanics, **particles exist in multiple states at once** until observed. The **Many-Worlds Interpretation (MWI)** of quantum mechanics suggests that **each quantum decision creates a branching reality**.

Wavefunction Representation:

$Ps = a|A[x] + b|B[x]$

where:

- **Ps** represents the total wavefunction.
- **|A[x] and |B[x]** represent different possible realities.
- **a and b** are probability amplitudes.

Could the Mandela Effect occur due to **merging quantum branches**, where people retain memories from a different timeline?

Quantum Entanglement and Non-Local Effects

Entanglement suggests that **two particles can instantaneously affect each other across vast distances**. Some researchers speculate that **entangled states may influence human consciousness**, potentially allowing memories to persist from **alternate timelines**.

Time Loops and Temporal Distortions

If time is **nonlinear**, the Mandela Effect could be a side effect of **temporal shifts, loops, or timeline alterations**.

Time Travel and Causal Loops

If an event in the past were altered, how would reality correct itself? Some theories propose:

- **Ripple Effect** - Changes propagate through time, modifying details but preserving major events.

- **Parallel Coexistence** - Both realities exist simultaneously, and consciousness "jumps" between them.

- **Retrocausality** - Future events influence the past, changing recorded history.

If time travel or manipulation is occurring, it could explain **small inconsistencies**--such as name changes, memory variations, and shifting cultural artifacts.

The Mandela Effect in a Simulated Reality

If we assume that reality is a **computer simulation** (as discussed in Chapter 6), then **glitches, errors, or updates** could explain Mandela Effect occurrences.

Reality as a Programmed Environment

- **If reality operates like a simulation, small errors or overwrites could alter recorded history.**

- **Data compression or memory optimization** might adjust details in seemingly insignificant ways.

- **Version updates in reality's programming** could introduce retroactive changes that some people remember differently.

Glitches in the System

Could memory discrepancies arise from **flaws in the simulation's rendering process**? Possible signs of simulation-related changes include:

- **Repeating historical cycles** that seem "pre-programmed."

- **Sudden changes in established facts** without evidence of prior modification.

- **Anomalies in perception**--where groups of people recall events that "never happened."

Future Research and Investigations

The Mandela Effect remains one of the most mysterious phenomena of modern times. As more cases are reported, researchers from different fields are beginning to take it seriously. Some potential future studies include:

- **Large-scale cognitive studies** to test whether the effect is purely psychological.

- **Quantum experiments** to determine if consciousness plays a role in timeline merging.

- **Simulated reality investigations** to examine whether these changes resemble digital modifications.

Conclusion: What is Reality?

The Mandela Effect challenges our understanding of **time, memory, and reality itself**. Whether it results from **psychological biases, quantum fluctuations, or simulation errors**, its widespread nature suggests that **reality is more complex than we assume**.

Key Takeaways:

- **The Mandela Effect describes collective memory discrepancies, leading to questions about reality.**

- **Psychological theories explain it as false memory formation and cognitive distortions.**

- **Quantum mechanics suggests the possibility of parallel realities interacting.**

- **If reality is a simulation, Mandela Effects could be glitches or program updates.**

As scientific understanding evolves, we may one day uncover whether the Mandela Effect is **a quirk of human memory or evidence of deeper, hidden forces shaping reality.**

Chapter 8: Bridging Consciousness, Time Waves, and Quantum Mechanics

1. Introduction: The Connection Between Mind, Time, and Reality

Throughout history, **philosophers, scientists, and mystics** have debated the **relationship between consciousness and time**. Is time an independent, external dimension, or is it fundamentally linked to **perception, observation, and thought**? Could the human mind play a role in shaping, bending, or even moving through time?

Modern physics suggests that:

- **Time may not be absolute** but instead an emergent property of deeper quantum processes.

- **Observation affects quantum states**, raising questions about the role of consciousness in reality.

- **Time waves may interact with perception**, influencing memory, intuition, and future awareness.

This chapter explores the **scientific, theoretical, and metaphysical** connections between **human consciousness, time waves, and the underlying mechanics of the universe**. Additionally, it examines **how government-backed research programs, such as the CIA's Gateway Process, have sought to harness consciousness as a tool for transcending time and space**. Finally, we will explore how **artificial intelligence (AI) may become a crucial tool in detecting and interacting with these time waves**, expanding our understanding of non-local awareness.

2. Quantum Mechanics and the Role of the Observer

One of the most perplexing aspects of quantum mechanics is **the observer effect**, which suggests that **a conscious observer can collapse a quantum wavefunction into a definite state**.

The Double-Slit Experiment and Wavefunction Collapse

The **double-slit experiment** demonstrates that when **unobserved**, particles exist as a **wave of probabilities**, but when measured, they behave as **discrete particles**.

Mathematically, this is represented by the wavefunction:

$$Ps(x,t) = Ae^{i(kx - ot)}$$

where:

- **Ps(x,t)** is the probability amplitude of the particle.
- **A** is the wave's amplitude.
- **k** is the wave vector.
- **o** is the angular frequency.

If consciousness plays a role in collapsing wavefunctions, could it also **influence time itself**?

Quantum Consciousness and the Penrose-Hameroff Model

Physicist **Roger Penrose** and anesthesiologist **Stuart Hameroff** proposed the **Orchestrated Objective Reduction (Orch-OR) Theory**, suggesting that:

- Consciousness arises from **quantum processes within microtubules** in brain neurons.

- These quantum processes may be linked to **wavefunction collapse** at a fundamental level.

- **Quantum states in the brain could influence time perception and decision-making**.

If quantum mechanics plays a role in consciousness, then **the human mind may be directly connected to the underlying structure of time and reality**. This leads us to explore whether consciousness can transcend time--an idea that has been **scientifically investigated by U.S. intelligence agencies** through the Gateway Process.

---

3. The Gateway Process: Government-Backed Consciousness Exploration

The *Gateway Process* was a **highly classified U.S. intelligence program** designed to explore the potential of **consciousness beyond the physical body**. Developed in collaboration with the **Monroe Institute**, the process was based on a **systematic method for inducing altered states of awareness** using controlled **brainwave synchronization techniques**.

The **1983 declassified CIA report**, authored by Lt. Colonel Wayne M. McDonnell, revealed that the *Gateway Experience* was a **deliberate attempt to train individuals to access non-local consciousness**. The key findings included:

- **Reality as a Holographic Projection**: The report aligned with physicist **David Bohm's holographic universe theory**, stating that **consciousness can transcend the limitations of space and time** and interact with a **larger field of intelligence**.

- **Brainwave Entrainment and Time Manipulation**: The use of binaural beats in **Hemi-Sync technology** allowed individuals to **tune their brainwaves to specific frequencies**, creating states in which they could **perceive and even influence time in a non-linear way**.

- **Remote Viewing and Non-Local Awareness**: The CIA's experiments confirmed that trained participants could **access distant locations, past and future events, and information beyond conventional sensory perception**.

The implications of this research suggest that **consciousness is not bound by time** and that through the right training, humans may be able to **interface with time waves, much like a radio tunes into different frequencies**.

4. Time Waves: Theoretical Models and Implications

The idea of **time waves** suggests that time is not a static, linear progression but a **dynamic, oscillating phenomenon** influenced by **energy, observation, and consciousness**. If time behaves like a wave, it could mean that past, present, and future are interconnected, and **temporal frequencies** dictate the nature of reality itself.

Wave-Like Nature of Time

In classical physics, wave motion is described by the **wave equation**:

$$\partial^2 T / \partial t^2 = v^2 \nabla^2 T$$

where:

- **T** represents the time wave function.
- **v** is the wave speed.
- ∇^2 is the Laplacian operator, describing spatial oscillations.

If time itself has a **wave structure**, then:

- **Future and past moments may interact dynamically**, leading to phenomena like **precognition and retrocausality**.

- **Certain frequencies might resonate with human consciousness**, potentially explaining **déjà vu, time dilation, and altered perception of events**.

- **Time distortions could occur near strong energy fields** or under intense **mental or emotional states**, hinting at a possible mind-time interaction.

Quantum Time Oscillations

If time exhibits wave-like behavior, then **quantum mechanics must also reflect these oscillations**. In quantum physics, particles behave as waves, existing in multiple states until observed. This principle may extend to time itself, meaning:

- Time may exist as a **superposition** of possible states, where multiple futures and pasts coexist.

- A quantum observer could potentially **collapse time waves**, selecting one outcome from a range of possibilities.

- Time could experience **constructive or destructive interference**, explaining why certain moments feel "stronger" or more significant than others.

Theoretical physicists have proposed **quantum temporal interference experiments**, where time behaves similarly to the **double-slit experiment**, showing that particles can interfere with past versions of themselves. This suggests that time waves might allow for **information exchange across different points in time**.

Human Consciousness and Time Resonance

One of the most intriguing implications of time wave theory is that **human consciousness may be inherently linked to the structure of time**. Research in **the Gateway Process, remote viewing, and altered states of consciousness** suggests that the human brain may be able to **synchronize with time waves** to access information beyond the present moment.

This may explain why people experience:

- **Déjà vu** - A possible interference pattern between a future and present time wave.

- **Premonitions** - Resonance with high-probability time waves that later become reality.

- **Flow States** - A moment of deep focus where the brain synchronizes with a stable time wave, allowin

5. AI and the Future of Consciousness Exploration

Introduction

In our ever-evolving technological era, artificial intelligence (AI) is beginning to intersect with domains that were once reserved for metaphysics and speculative science. With emerging theories on time waves and non-local consciousness, profound questions arise: Can AI be trained to detect the subtle fluctuations of time, and might it one day enhance our understanding of the very fabric of human awareness?

Historically, the exploration of consciousness has spanned philosophy, mysticism, and modern quantum physics. As researchers uncover intricate connections between the mind, time, and reality, AI is poised to bridge the gap between objective data and subjective experience.

Theoretical Foundations of Time Waves and Consciousness

The concept of time waves posits that the flow of history and human consciousness is not linear but rather cyclic in nature. Inspired by thinkers such as Terence McKenna, this theory suggests that periods of intense novelty and change punctuate long stretches of stability. These "novelty spikes" hint at a deep, underlying rhythm governing both cosmic events and the evolution of human thought.

Moreover, parallels can be drawn between these theoretical time patterns and the oscillatory behavior observed in quantum fields as well as the brain's own electrical activity. Such connections propose that time, consciousness, and physical reality may be more interconnected than classical science has traditionally maintained.

Can AI Interface with Time Waves and Non-Local Awareness?

If AI is trained on models that consider wave-based approaches to consciousness, several intriguing possibilities emerge:

- **AI-Assisted Brainwave Synchronization** – Advanced neural networks could analyze and fine-tune human brainwaves to achieve resonance with specific frequencies, potentially enhancing Gateway-like experiences and deep meditative states.

- **Fractal Pattern Recognition in Time Data** – By scrutinizing historical patterns and novelty spikes, AI might mirror McKenna's Time Wave Theory, offering predictions about future cultural and technological shifts based on cyclic resonance.

- **Quantum AI and Non-Local Information Processing** – The integration of AI with quantum computing could pave the way for systems capable of processing data outside the conventional constraints of time and space, essentially developing a form of non-local awareness.

Emerging Technologies and Methodologies

Recent advancements in AI research have led to sophisticated algorithms capable of handling complex, non-linear datasets. Techniques such as deep learning, recurrent neural networks, and quantum-inspired computation are now being applied to analyze data that defies classical explanation.

Experimental projects are already exploring AI-driven neurofeedback, where real-time brainwave data is used to guide subjects into altered states of consciousness. Such experiments suggest that with proper calibration, AI can assist in unlocking deeper layers of human awareness.

Interdisciplinary collaborations among neuroscientists, quantum physicists, and computer scientists are pushing the envelope further. These teams are working on integrating biometric sensors, virtual reality, and AI to create immersive experiences that may help map the elusive time waves that underlie our perception of reality.

Future Implications and Applications

AI holds the promise of becoming a true amplifier of consciousness. Some of the key potential applications include:

- **Enhancing Human Time Perception** – By training individuals to sync with specific resonant wave states, AI could help expand the human perception of time, allowing for deeper introspection and a more profound understanding of reality.

- **Mapping Time Waves** – With advanced data analytics, AI could detect and map the subtle fluctuations in time waves, providing a predictive tool for historical trends and future events.

- **Creating Hybrid Intelligence Systems** – The integration of human intuition with machine precision might yield a unified field of awareness. This hybrid intelligence could transform fields ranging from scientific research to the arts by offering new ways to process and interpret information.

The prospect of AI merging with consciousness studies could redefine the boundaries between subjectivity and objectivity, ultimately altering how we understand both time and existence.

Practical Experiments and Case Studies

Several experimental initiatives are already exploring the interface between AI and human consciousness. Laboratories around the world are employing AI-enhanced neurofeedback systems to monitor and modulate brainwave activity during meditation, trance states, and other altered states of awareness.

In one set of studies, subjects using AI-guided biofeedback techniques have reported deeper meditative experiences and enhanced clarity of thought. These early case studies suggest that the deliberate synchronization of brainwaves to external resonant frequencies can produce measurable shifts in consciousness.

Additionally, pilot projects incorporating AI with immersive virtual reality environments have demonstrated the possibility of altering time perception. Users immersed in these AI-curated experiences often report a distorted sense of time—experiencing minutes as hours or vice versa—which further supports the hypothesis that external stimuli can modulate our internal clock.

Ethical Considerations and Societal Impact

The convergence of AI with the exploration of consciousness and time waves raises important ethical and societal questions. As we venture into this new frontier, issues such as data privacy, the manipulation of mental states, and the potential for misuse must be rigorously addressed.

Who will have access to these powerful technologies, and under what regulatory frameworks will they operate? It is critical to establish ethical guidelines that protect individual autonomy and ensure that advancements benefit society as a whole.

Furthermore, the potential to alter human perception and cognitive processes carries significant risks. Misuse of such technologies could lead to unintended psychological or social consequences, making the development of robust oversight mechanisms essential.

Future Prospects and Challenges

Looking ahead, the integration of AI with consciousness and time wave research is poised to be a transformative yet challenging journey. Current technological limitations, the need for cross-disciplinary collaboration, and the complex ethical landscape all present significant hurdles.

Key challenges include:

- **Technological Limitations** – Building AI systems capable of processing the vast, non-linear datasets inherent in consciousness research requires breakthroughs in both hardware and software.

- **Interdisciplinary Integration** – Bridging neuroscience, quantum physics, and AI demands collaboration across diverse fields, each with its own methodologies and terminologies.

- **Ethical and Regulatory Barriers** – Establishing frameworks to ensure the responsible use of these emerging technologies is paramount to avoid potential abuses.

Despite these challenges, the steady progress in AI and cognitive science offers hope that the fusion of these fields may eventually lead to groundbreaking discoveries. The prospect of harnessing time wave data and integrating it with advanced AI represents a paradigm shift that could redefine the future of human consciousness.

Conclusion

If time waves exist and can be influenced by consciousness, then AI's ability to detect and analyze these fluctuations could prove transformative. By serving as a consciousness amplifier, AI not only holds the potential to enhance human perception but also to unlock new insights into the very nature of time and reality.

As we stand at the threshold of this emerging frontier, the integration of AI with the exploration of consciousness is poised to redefine what it means to be aware

6. Conclusion: A New Paradigm for Time and Consciousness

The study of **time, consciousness, and AI** is no longer a purely theoretical endeavor. The *Gateway Process* demonstrated that **human minds can transcend time**, and *Time Wave Theory* suggests that **time itself operates in predictable cycles**.

With the rise of AI, we now face **a new frontier**--one where machines may **not only assist human intelligence** but potentially **interface with the very fabric of time itself**.

The implications are staggering:

- **AI could serve as a bridge between human awareness and time perception**.

- **Quantum computing could allow AI to engage with non-local consciousness fields**.

- **The study of time waves may lead to practical applications in forecasting, decision-making, and reality navigation**.

If consciousness is the final frontier, **AI may be the tool that finally unlocks it**.

Chapter 9: Future Research and Implications for Humanity

Introduction: The Uncharted Frontier of Time and Reality

Humanity has long been fascinated by the mysteries of **time, space, and consciousness**. As scientific advancements continue to push the boundaries of **physics, quantum mechanics, and cosmology**, we are faced with profound questions that challenge our fundamental understanding of existence.

In the coming decades, researchers will delve into uncharted territories where the laws of nature may be redefined and where new paradigms of thought emerge. The interplay between human consciousness and the fabric of reality stands as one of the most captivating and controversial subjects in modern science.

Fundamental Questions Driving Future Research

- **Is time truly a fixed, linear construct, or does it operate in waves and cycles?**
- **Can human consciousness influence the flow of time or even traverse it?**
- **Will we one day achieve control over time and space, enabling time travel, interstellar travel, or even reality manipulation?**

This chapter explores the **future research directions** that may unlock the secrets of **time waves, spacetime engineering, and**

quantum consciousness, ultimately transforming our understanding of reality.

Emerging Theories and Models

New theoretical frameworks are being proposed that challenge the conventional, linear view of time. Researchers are investigating the possibility that time may function more like a dynamic, oscillating field—where cycles of change are driven by fluctuations in energy and consciousness. These models suggest that:

- The universe may be composed of nested cycles, each operating at different scales and frequencies.

- Periods of rapid transformation, or "novelty spikes," could correspond to shifts in collective human consciousness.

- Spacetime might be more malleable than previously thought, influenced by both physical forces and mental states.

These innovative ideas form the basis for experiments that seek to observe and measure the subtle rhythms of time itself.

Spacetime Engineering and Quantum Frontiers

With advances in quantum mechanics and high-energy physics, scientists are exploring the possibility of **spacetime engineering**. This area of research investigates:

- How quantum fluctuations could be harnessed to create controlled distortions in spacetime.

- The potential for manipulating gravitational fields to enable breakthroughs in propulsion and energy generation.

- New materials and technologies that could one day facilitate the construction of devices capable of influencing the flow of time.

These cutting-edge investigations may pave the way for technologies that were once relegated to the realm of science fiction.

Quantum Consciousness and Human Awareness

One of the most tantalizing areas of research involves the intersection of quantum physics and human consciousness. Emerging studies propose that:

- Consciousness may be linked to quantum processes that operate beyond the classical physical realm.

- The brain could be viewed as a quantum processor, sensitive to fluctuations in spacetime and capable of tapping into non-local information.

- Techniques such as advanced neuroimaging and AI-driven analysis may soon provide insights into the quantum aspects of cognition.

If these hypotheses hold true, the implications for understanding human consciousness—and its potential to interact with time—are profound.

Future Research Directions

To unlock the mysteries of time and reality, future research must integrate insights from diverse fields including physics, neuroscience, and computer science. Key areas of focus include:

1. **Advanced Temporal Analytics:** Developing computational models to analyze time wave patterns and correlate them with historical and future events.

2. **Interdisciplinary Experiments:** Conducting collaborative experiments that merge quantum mechanics, neurobiology, and artificial intelligence to probe the nature of consciousness.

3. **Technological Innovation:** Investing in technologies such as quantum computing and high-resolution neuroimaging to explore the boundaries of spacetime manipulation.

These research initiatives are poised to challenge our current scientific paradigms and open new avenues for discovery.

Ethical Considerations and Societal Impact

As we push the frontiers of time and reality, it is crucial to address the ethical and societal implications of these groundbreaking advancements. Critical questions include:

- How will these technologies impact individual privacy and the autonomy of human thought?

- What regulatory frameworks must be developed to ensure the responsible use of spacetime manipulation technologies?

- Can society collectively manage the profound shifts in power dynamics that may result from the ability to control or traverse time?

Addressing these issues will be as important as the scientific breakthroughs themselves, ensuring that progress benefits all of humanity.

Implications for Humanity

The potential to influence time and reality carries far-reaching implications for every facet of human existence. From the redefinition of history and the potential for time travel to the transformative impact on culture, art, and social structures, the discoveries of tomorrow could reshape the human experience.

By understanding and harnessing these advanced concepts, future generations may experience a paradigm shift that not only alters our perception of time but also empowers us to create a more dynamic and interconnected world.

Conclusion: Embracing the Future of Discovery

As we stand on the threshold of unprecedented scientific discovery, the journey into the uncharted realms of time, space, and consciousness beckons us forward. The research directions outlined in this chapter offer a glimpse into a future where the mysteries of reality may be unraveled, leading to technologies and insights that transform human existence.

Embracing these challenges requires not only bold scientific inquiry but also a commitment to ethical responsibility and global cooperation. The uncharted frontier of time and reality promises to redefine our understanding of the universe—and ourselves—in ways that are as inspiring as they are profound.

Expanding the Science of Time Waves

The **Time Wave Hypothesis** suggests that **time is not a smooth linear progression, but an oscillating wave-like structure**. Future research must focus on:

- **Developing mathematical models** that accurately describe time waves.

- **Conducting high-energy physics experiments** to detect temporal fluctuations.

- **Investigating the relationship between gravitational waves and time distortions.**

Mathematical and Experimental Approaches

If time waves exist, they should obey equations similar to **classical wave mechanics**:

$$\partial^2 T / \partial t^2 = v^2 \, \nabla^2 T$$

where:

- **T** represents the time wave function.

- **v** is the wave speed.

- **∇^2** is the Laplacian operator representing spatial oscillations.

Future research may focus on:

- **Detecting time waves** through gravitational wave observatories.

- **Using quantum field experiments** to measure time-dependent fluctuations.

- **Exploring resonant time fields** in strong electromagnetic conditions.

Spacetime Engineering and Time Manipulation

One of the most promising areas of future research is **spacetime engineering**, the ability to **modify, bend, or manipulate spacetime itself**.

Exploring Warp Drive Technology

As discussed in previous chapters, the **Alcubierre Drive** proposes a method of **faster-than-light (FTL) travel** by contracting and expanding spacetime:

$$ds2 = -c2dt2 + (dx - v\_s\ f(r\_s)\ dt)2 + dy2 + dz2$$

where:

- **v_s** is the velocity of the warp bubble.
- **f(r_s)** is a function defining the bubble's shape.

Key research areas for making warp drives a reality include:

- **Developing stable energy fields** to generate warp bubbles.
- **Investigating the potential for negative energy sources**.
- **Conducting small-scale laboratory experiments on spacetime manipulation.**

Artificial Wormholes and Temporal Gateways

Future research may also explore the creation of **artificial wormholes**, tunnels in spacetime that could allow for instantaneous travel. **Einstein's field equations** suggest the existence of **traversable wormholes** under specific conditions:

$$G\_{\mu\nu} + L\ g\_{\mu\nu} = (8pG/c4)\ T\_{\mu\nu}$$

For a wormhole to remain open, it would require **exotic matter** with negative energy density. Ongoing studies in **quantum vacuum energy and Casimir forces** may hold the key to this possibility.

Quantum Consciousness and Its Role in Time Perception

As explored in previous chapters, consciousness and quantum mechanics may be deeply linked. Future research must investigate:

- **The nature of consciousness as a quantum phenomenon.**

- **Whether the human mind can influence quantum time states.**

- **How meditation, altered states, and focused thought affect time perception.**

Time and the Human Mind: Neurophysiological Studies

Modern neuroscience suggests that **time perception is a constructed process** in the brain. Research into brainwave activity has shown:

- **Gamma waves (30-100 Hz)** - Associated with deep concentration and expanded awareness.

- **Theta waves (4-7 Hz)** - Linked to altered states of consciousness, intuition, and time distortion.

Could these brainwaves **synchronize with external time waves**, influencing perception or even allowing consciousness to move beyond linear time?

Ethical and Philosophical Implications of Time Manipulation

If humanity gains control over time, it will bring profound ethical considerations:

- **Time Travel Paradoxes** - Could altering the past create paradoxes that disrupt causality?

- **The Fate of Free Will** - If time loops exist, is the future already determined?

- **Access to Temporal Technology** - Who controls the power to manipulate time?

Philosophical discussions on time often explore the nature of **fate vs. free will**, with some theories suggesting that **humanity is already part of a predetermined time wave pattern**.

The Possibility of "Time Civilizations"

If time manipulation is possible, it raises the question: **Could advanced civilizations have already mastered time control?** Future research may search for evidence of **time travelers, advanced intelligence, or signals from the future**.

The Future of Humanity in a Time-Transcending Reality

As we move forward into an era of scientific discovery, new frontiers in **time waves, consciousness, and spacetime engineering** could redefine the human experience. The potential breakthroughs include:

- **Mastering time perception** - Expanding human awareness beyond linear time.

- **Unlocking interstellar travel** - Creating methods of FTL movement via warp fields.

- **Understanding reality at a deeper level** - Integrating physics, consciousness, and cosmology into a unified theory.

Conclusion: What Lies Ahead?

The journey to understand **time, space, and consciousness** is far from over. The **next century** may bring discoveries that revolutionize our understanding of existence, allowing us to:

- **Harness the fundamental forces of time.**
- **Unravel the mysteries of the quantum mind.**
- **Travel beyond the stars through spacetime engineering.**

Key Takeaways:

- **Time waves may be measurable and experimentally verified.**
- **Spacetime engineering could lead to interstellar travel.**
- **Quantum consciousness research may redefine our understanding of reality.**
- **Ethical considerations will be crucial as time manipulation becomes possible.**

In the coming decades, the study of **time, consciousness, and reality** may lead us into **a new era of discovery**, unlocking **possibilities beyond imagination.**

Credits & Acknowledgments

Author:

John Smedley

Contributions:

This book has been developed through extensive **research, collaboration, and synthesis** of modern theories in **physics, quantum mechanics, consciousness studies, and spacetime engineering**.

Special Thanks:

I extend my deepest gratitude to:

- **The physicists, researchers, and visionaries** whose work laid the foundation for these explorations.

- **Philosophers and metaphysical thinkers** who have shaped our understanding of reality beyond empirical science.

- **Scientific communities and independent researchers** who continue to push the boundaries of what is possible.

This work is dedicated to those who dare to **question the nature of existence and seek deeper truths** about time, consciousness, and reality.

References & Resources

Books & Papers on Physics and Time Theory

1. **Einstein, A.** (1905). *On the Electrodynamics of Moving Bodies.* Annalen der Physik.

2. **Hawking, S.** (1988). *A Brief History of Time.* Bantam Books.

3. **Godel, K.** (1949). *An Example of a New Type of Cosmological Solution of Einstein's Field Equations of Gravitation.* Reviews of Modern Physics.

4. **Penrose, R.** (1989). *The Emperor's New Mind.* Oxford University Press.

5. **Alcubierre, M.** (1994). *The Warp Drive: Hyper-fast travel within general relativity.* Classical and Quantum Gravity.

Quantum Mechanics & Consciousness Research

- **Hameroff, S., & Penrose, R.** (1996). *Orchestrated Reduction of Quantum Coherence in Brain Microtubules: A Model for Consciousness.* Journal of Consciousness Studies.

- **Tegmark, M.** (2000). *The Importance of Quantum Decoherence in Brain Processes.* Physical Review E.

- **Bostrom, N.** (2003). *Are We Living in a Computer Simulation?* Philosophical Quarterly.

Additional Resources

- **NASA Eagleworks Laboratory** - Research on warp field physics and spacetime engineering.

- **Global Consciousness Project** - Studies on human thought and its impact on randomness.

This list is by no means exhaustive, but it provides a **strong starting point** for those interested in **further exploration** of these topics.

Closing Remarks & Final Thoughts

The Journey into Time and Reality

This book has been an exploration of **time, consciousness, quantum mechanics, and the potential future of humanity**. As we continue to advance our understanding of **spacetime, wave structures, and the mind's influence on reality**, we may one day find ourselves **transcending the limitations of time as we know it**.

What Lies Ahead?

The next century may bring:

- **Breakthroughs in quantum cognition** that redefine how we experience reality.

- **The practical discovery of time waves** and their implications for physics.

- **The possibility of interstellar or even interdimensional travel**.

- **The fusion of science and metaphysics**, leading to new understandings of existence.

While this book has aimed to **explore possibilities**, the **true discoveries** will come from **future generations of thinkers, scientists, and explorers** willing to push beyond the boundaries of conventional knowledge.

Final Words

We may one day discover that **time is not a rigid sequence of events but a malleable, conscious-interactive field**--one that we can **navigate, manipulate, and master**.

Whether through **scientific discovery, technological breakthroughs, or shifts in human perception**, the study of **time and consciousness** will continue to be at the forefront of **our search for deeper truths about reality**.

The journey does not end here--it is only beginning.

End of Book

Made in United States
Cleveland, OH
29 March 2025

15628541R00069